Australian Snakes in My Backyard

By James Toohill

First Edition
September, 2017
ISBN 9781977916389

© James Toohill 2017

Toohill, James 2005 –

This book is copyright. Apart from any fair dealing for the purpose of private study, research, criticism or review, as permitted under the Copyright Act, no part may be reproduced by any process without written permission. Enquiries should be addressed to the Publishers.

All rights reserved.

Published by Wild Colonial Press
PO Box 7151
Mount Crosby Qld 4306
Australia

jerilderierareandcollectablebooks@live.com.au

First published 2017

Me and a small carpet python found on my property.

For

My Family

&

My Dog Mittens

ACKNOWLEDGEMENTS

A huge thank you to the Queensland Museum for allowing me the use of many of the amazing snake photographs in this book.

CONTENTS

Page	5	Dedication
Page	7	Acknowledgements
Page	9	Contents
Page	11	Introduction
Page	13-14	Green Tree Snake
Page	15-17	Red-Bellied Black Snake
Page	19-23	Carpet Python
Page	25-26	Brown Tree Snake
Page	27-28	Eastern Brown Snake
Page	29-31	Yellow-faced Whip Snake
Page	33-35	Common Death Adder

INTRODUCTION

G'day, fellow people of the bush! My name is James Toohill and I am 11 years old. I wrote this book because of my passion and love for Australian snakes. When I was two years old, I was bitten on the hand by a red-bellied black snake in my backyard. LUCKILY I survived, because it was a dry bite. Snakes don't always inject venom when they bite. Sometimes they give a warning bite, which is what happened to me. I still have a scar on my hand to this day, or as I like to call it 'a badge of courage'. Even though I was bitten, I still love snakes, and want to help educate other people on how amazing and important to the environment Australian snakes are.

This book contains interesting facts about some of the different types of Australian snakes I have found in my backyard, which is located in the western region of Brisbane, Queensland, Australia.

GREEN TREE SNAKE

(Photo courtesy of The Queensland Museum)

Question:- What colour is a green tree snake?

Answer:- The green tree snake is usually green, but may be black, blue, or yellow with pale blue flecks, with a green or bright yellow throat.

Question:- How long does a green tree snake get?

Answer:- A green tree snake grows up to 2 metres long.

Question:- Where do you find green tree snakes?

Answer:- You find green tree snakes in Northern and Eastern Australia.

Question:- Where does a green tree snake live?

Answer:- A green tree snake lives in gardens, rural lands and rainforest habitats.

Question:- When is a green tree snake active?

Answer:- A green tree snake is active during the day.

Question:- Is a green tree snake venomous or non-venomous?

Answer:- A green tree snake is non-venomous, which means it is not poisonous.

Question:- What does a green tree snake eat?

Answer:- A green tree snake eats frogs, fish and small reptiles.

Question:- How many eggs does a green tree snake lay?

Answer:- A green tree snake lays 3 to 16 eggs.

Question:- How long is a baby green tree snake when it hatches from its egg?

Answer:- A baby green tree snake is 24cm long when it hatches from its egg.

Question:- Are their other similar species to the green tree snake?

Answer:- There are no similar species to the green tree snake.

RED-BELLIED BLACK SNAKE

Pictured Below:- A red-bellied black snake slithering through the grass.

Question:- What does a red-bellied black snake look like?

Answer:- A red-bellied black snake has a glossy black back and a red underbelly, which makes it easy to identify.

Question:- How long does a red-bellied black snake grow?

Answer:- A red-bellied black snake will grow up to 2 metres in length.

Question:- Where are red-bellied black snakes found?

Answer:- A red-bellied black snake is found in north-eastern Queensland, south-eastern Queensland, New South Wales and Victoria. (I even found one in my lounge-room!!!!)

Question:- What habitat do red-bellied black snakes like?

Answer:- Red-bellied black snakes are found in well-watered areas, such as near rivers, creek banks, swamps and rainforests.

Question:- When are red-bellied black snakes active?

Answer:- Red-bellied black snakes are active during the daytime.

(Photo courtesy of The Queensland Museum)

Question:- Are red-bellied black snakes venomous or dangerous?

Answer:- Red-bellied black snakes are venomous and dangerous, if bitten seek medical attention as quickly as possible.

Question:- What do red-bellied black snakes eat?

Answer:- Red-bellied black snakes feed on fish, frogs, other snakes, reptiles and small mammals. They may also eat cane toads, however this results in the snake being poisoned by the cane toad, and the red-bellied black snake will die.

Question:- How many babies does a red-bellied black snake have?

Answer:- A red-bellied black snake will have 5 to 19 live young, born between October and March.

Question:- How big are red-bellied black snake babies when they are born?

Answer:- Red-bellied black snake babies are 22cm long, when they are born.

(Photo courtesy of The Queensland Museum)

CARPET PYTHON

Pictured Below:- A carpet python slithering around a backyard

Question:- How do you identify a carpet python?

Answer:- A carpet python is very easy to identify due to their white underbelly and white, black and brown splotches.

Question:- How long can a carpet python grow?

Answer:- A carpet python can grow up to 3 metres in length.

Question:- Where are carpet pythons found?

Answer:- Carpet pythons are found in the northern, eastern and southern areas of Australia.

*Pictured Above – This carpet python was hunting at night in my front **yard***

Question:- Where does a carpet python live?

Answer:- A carpet python lives in open forests, rainforests, rural lands, parklands, and suburban gardens / your garden!!! It can also be found in trees, on the ground, or in buildings / houses.

Question:- When are carpet pythons active?

Answer:- Carpet pythons are active by day and night.

Question:- Are carpet pythons venomous?

Answer:- Carpet pythons are non-venomous, but can give a very nasty bite if harmed, which can cause infections.

Question:- What kind of teeth does a carpet python have?

Answer:- A carpet python has razor-like teeth, not fangs.

Question:- What does a carpet python eat?

Answer:- A carpet python eats frogs, lizards, birds and small mammals, but will also eat cane toads, with a deadly consequence.

Question:- How many babies does a carpet python have?

Answer:- A carpet python, in summer, lays 10 to 47 eggs.

Question:- How does a mother carpet python keep her eggs warm?

Answer:- A mother carpet python will shiver to create heat to keep her eggs warm.

(Photo courtesy of The Queensland Museum)

Question:- Is a carpet python mother protective over her eggs?

Answer:- Yes, a mother carpet python is protective of her eggs.

Question:- How big are baby carpet pythons when they are born?

Answer:- Baby carpet pythons are around 39cm long when they are born, (longer than a ruler).

Pictured Above – Me and a small carpet python, which I found wandering around my house.

BROWN TREE SNAKE

(Photo courtesy of The Queensland Museum)

Question:- What does a brown tree snake's head look like?

Answer:- A brown tree snake has a large head, vertical pupils, big eyes and a narrow head.

Question:- What colouration is a brown tree snake?

Answer:- A brown tree snake has a kind of zig-zag pattern of dark brown and light brown.

Question:- How long does a brown tree snake grow?

Answer:- A brown tree snake grows up to 2 metres in length.

Question:- Where are brown tree snakes found?

Answer:- Brown tree snakes are found in northern, eastern and southern Australia, down to the Sydney region. They are even often found curled up in buildings.

Question:- When are brown tree snakes active?

Answer:- Brown tree snakes are active at night.

Question:- Are brown tree snakes venomous?

Answer:- Brown tree snakes are venomous. They are not considered overly dangerous, but if bitten seek immediate medical attention.

Question:- What does a brown tree snake eat?

Answer:- Brown tree snakes eat birds and their eggs, frogs and mammals. They don't eat reptile eggs.

Question:- Do any other snakes look like a brown tree snake?

Answer:- There are no other species that look like the brown tree snake.

Question:- How many eggs does a brown tree snake lay?

Answer:- A brown tree snake lays 3 – 11 eggs. When the eggs hatch they are around 30cm long.

EASTERN BROWN SNAKE

Question:- How long can an eastern brown snake get?

Answer:- An eastern brown snake can grow up to 2.2 metres in length.

Question:- What colour is an eastern brown snake?

Answer:- Eastern brown snakes colour can vary a lot from brown, orange, russet to almost black. It can be one plain colour or banded.

Question:- What colour is an eastern brown snakes underbelly?

Answer:- An eastern brown snakes underbelly is usually cream with orange spots.

Question:- Are eastern brown snakes venomous?

Answer:- Eastern brown snakes are highly venomous and if bitten you should seek urgent immediate medical attention.

Question:- Are eastern brown snakes found in South-east Queensland?

Answer:- Yes, eastern brown snakes are found all over south-east Queensland.

Question:- Do eastern brown snakes look like other snakes?

Answer:- Yes, eastern brown snakes can be confused with a number of other snakes from the coastal taipan, the tiger snake, the mulga snake, some western brown snakes, the red-naped snake, the grey snake, the dwyer's snake and the curl snake.

Question:- What habitat do eastern brown snakes like?

Answer:- Eastern brown snakes like to live in dry open forests or in disturbed agricultural lands.

Question:- What does an eastern brown snake like to eat?

Answer:- An eastern brown snake likes to eat lizards, mammals, frogs, birds, snakes, and reptile eggs.

Question:- How many eggs does an eastern brown snake lay?

Answer:- An eastern brown snake lays up to 28 eggs per clutch.

(Photo courtesy of The Queensland Museum)

YELLOW FACED WHIP SNAKE

Pictured Below:- Yellow-faced Whip Snake slithering through the grass.

Question:- What does a yellow-faced whip snake look like?

Answer:- A yellow-faced whip snake has a very slender body, a black comma-shaped marking beneath its eye and on either side of this black marking is a yellow thin lined marking. It has a pale bluish grey to light olive green coloured body with a greenish grey coloured underbelly.

Question:- Where are yellow-faced whip snakes found?

Answer:- Yellow-faced whip snakes are found over most of mainland Australia.

Question:- What does a yellow-faced whip snake eat?

Answer:- A yellow-faced whip snake eats lizards and their eggs, frogs and other snakes.

Question:- Is a yellow-faced whip snake venomous?

Answer:- A yellow-faced whip snake is venomous. If bitten you should seek immediate medical attention.

(Photo courtesy of The Queensland Museum)

Question:- When are yellow-faced whip snakes active?

Answer:- Yellow-faced whip snakes are active by day, they are fast and alert.

(photo courtesy of The Queensland Museum)

Question:- How many eggs does a yellow-faced whip snake lay?

Answer:- A yellow-faced whip snake lays 9 eggs during February and March.

Question:- How big are baby yellow-faced whip snakes when they are born?

Answer:- Yellow-faced whip snakes are approximately 17cm long when they are born.

Question:- Are there any similar snake species to a yellow-faced whip snake?

Answer:- Yes, a green tree snake is similar to the yellow-faced whip snake, however, a green tree snake lacks the markings around the eyes of the yellow-faced whip snake.

COMMON DEATH ADDER

(Photo courtesy of The Queensland Museum)

Question:- What does a death adder look like?

Answer:- A death adder has a stocky body with an arrow-shaped head. It has a thin tail tip and ends with a short spine.

Question:- What colour is a death adder?

Answer:- A death adders back can be shades of grey to reddish-brown, and is usually marked with lighter coloured bands. It has a belly which is greyish to cream.

Question:- Where are death adders found?

Answer:- A death adder is found in eastern Australia, but not in far north and south), and southern South Australia and Western Australia.

Question:- What habitat does a death adder like?

Answer:- A death adder lives in wet and dry eucalypt forests, woodlands and coastal heaths.

Question:- Is a death adder good at hiding?

Answer:- Yes, a death adder likes to sit very still, hiding in leaf litter.

Question:- Are death adders venomous?

Answer:- Death adders are very venomous. If bitten seek immediate urgent medical attention.

Question:- What does a death adder eat?

Answer:- A death adder likes to eat small reptiles, birds, mammals, frogs. Sometimes death adders eat cane toads, but this will end up in death for the snake.

(Photo courtesy of The Queensland Museum)

Question:- What does a death adder use its tail for?

Answer:- A death adder uses its tail tip to attract prey by tricking the prey to come close, to eat, and then it strikes.

Question:- Does a death adder lay eggs or have live babies?

Answer:- A death adder has live babies.

Question:- How many babies does a death adder have?

Answer:- A death adder has up to 42 babies, between December and March.

Question:- What size are death adder babies when they are born?

Answer:- Death adder babies are around 12cm long when they are born.

Question:- Are there any similar species to the death adder?

Answer:- The De Vis' Banded Snake, and the Ornamental Snake look a little bit like a death adder.

Question:- Are there multiple species of death adder?

Answer:- Yes, there are multiple species of death adder, and all should be treated with extreme caution.

The End

www.ingramcontent.com/pod-product-compliance
Lightning Source LLC
Chambersburg PA
CBHW040250220526
45473CB00001B/434